Alfred Kirstein, Max Thorner

Autoscopy of the larynx and the trachea

Alfred Kirstein, Max Thorner

Autoscopy of the larynx and the trachea

ISBN/EAN: 9783744738316

Printed in Europe, USA, Canada, Australia, Japan

Cover: Foto ©berggeist007 / pixelio.de

More available books at **www.hansebooks.com**

THE

BANKING SYSTEM

OF THE

UNITED STATES

AND

ITS RELATION TO THE MONEY AND BUSINESS OF
THE COUNTRY.

BY

CHARLES G. DAWES.

1894:
RAND, McNALLY & COMPANY,
CHICAGO.

PREFACE.

The exact relations of our banking system to the business prosperity and money of the United States do not seem to be fully comprehended by our people.

The writer is a business man, to whom his dealings with banks, in the ordinary course of business, has suggested the preparation of this little volume, in which he discusses that service of banks to the community which to him seems at once the most important, and, at the same time, most commonly ignored. In his judgment there can be no proper understanding of the monetary problems of the day unless the relation of the bank-credit money of the country to the money of the Government is fully comprehended.

LINCOLN, NEB., October 29, 1894.

BANKING SYSTEM OF THE UNITED STATES.

CHAPTER I.

THE BANK AND THE COMMUNITY.

From the standpoint from which we shall consider a bank in its relation to the business community, we shall not treat it as a private corporation organized for profit, but will regard it as in the nature of a public corporation, its officers as public officers serving the business community, its depositors, whose funds are loaned by these officers to the community, as demand creditors of the community, and its borrowers as debtors of the depositors. From this standpoint the capital of a bank may be regarded as a kind of indemnity money, deposited by the stockholders of the bank as a pledge that the depositors in the community shall collect from the borrowers in the community " on demand."

As to the profits arising to the banking corporation from this species of trusteeship for the community, they are smaller than those of any other class of private corporations which deal with the community as a whole, and which perform services for the community which can be compared in value with those performed by the banks. The tax upon the community represented by these profits is so small, in comparison with the benefits derived by the community from banks, and compared with the risks incurred by banks for the sake of the community, that it is not worthy of extended consideration. The object of this little volume is to demonstrate the value and the vast importance of banks as the creators of the great bulk of the money used by the business community, and to point out some of the results arising out of the relations existing between our banking system and the business community, which are often overlooked. Before taking these matters up we will mention a few of the better known services of the bank to the community.

In the first place, a bank acts as an agent for the active and enterprising members of the community who borrow money for the sake of carrying on the productive and mercantile industries upon which so many members of every community depend for their living. It collects a multitude of small amounts from the members of the community, which, if retained by them, would be idle capital, and, under proper restrictions, places the money in the hands of those who put it to its best use. It acts as the trustee for the accumulating and saving portion of the community, affording them the opportunity to place their money where it is safe, and where at any time, unless otherwise contracted, they can receive it on demand.

By the use of a clearing-house and the check system, banks enable a community to carry on the bulk of their business without the risk and trouble of handling actual money in large sums. In returning the canceled checks to a customer, and balancing monthly his bank book, they furnish him with vouchers for payments made by him, signed by the party receiving the payments. To quite a portion of the community they act as bookkeepers, and among smaller tradesmen, and some larger ones, unfortunately, the stubs of the bank-check book form their most reliable accounts.

They become trustees in a multitude of transfers between parties residing in different places where papers are to be delivered on receipt of cash for remittance. They furnish drafts and bills of exchange, enabling the members of a community to cheaply and safely transfer money to a distance. They furnish information to business men, at home and at a distance, as to the reliability of prospective customers. Their favorable introductions facilitate business acquaintances. As National banks they add to the circulating medium "National Bank Notes." These are some of the services performed by banks for the community. But we are yet to consider their most important service.

CHAPTER II.

THE BANK-CREDIT MONEY OF THE UNITED STATES.

According to the report of the Secretary of the Treasury there is outstanding money issued by the United States Government as follows:

	General Stock Coined or Issued.	In Treasury.	Amount in Circulation Oct. 1, 1894.
Gold coin	$579,728,587	$79,602,339	$500,126,248
Standard silver dollars	421,176,408	366,900,165	54,276,243
Subsidiary silver	75,054,481	16,809,713	58,244,768
Gold certificates	64,845,699	55,260	64,790,439
Silver certificates	339,676,504	9,155,785	330,520,719
Treasury notes (1890)	151,609,267	30,113,893	121,495,374
United States notes	346,681,016	79,397,535	267,283,481
Currency certificates (1872)	56,305,000	550,000	55,755,000
National bank notes	207,564,458	5,017,748	202,546,710
Totals	$2,242,641,420	$587,602,438	$1,655,038,982

At the present time all of the money of the Government circulates at par. In other words, there is at present no form of money at a discount. Considered from the standpoint of intrinsic value alone, the gold dollar is of course the most valuable, although its value as money is no more than that of the paper or silver dollar. There seems to be, among our people, some confusion in ideas as to what this parity in the purchasing power of these different forms of money is due. Some possess the erroneous idea that the "stamp of Government" determines the matter, and that paper to which the "stamp of Government" is affixed, becomes *ipso facto* as valuable as the piece of gold bearing the same stamp.

It is well to fix clearly in our minds the true reason why there is no disparity in the debt-paying and purchasing power of the different forms of Government money.

The reason that every silver or paper dollar, issued by our Government, has the buying and debt-paying value of a gold dollar, is because it has been declared by law to be the established policy of the United

States to keep the metals, silver and gold, on a parity, and because, in the fulfillment of this policy, the Treasury Department of the Government holds itself in readiness to redeem, and in actual practice does redeem, in gold, upon demand, other forms of United States money in circulation, including paper and silver.

The amount of gold which has been held in the Treasury, and which has been practically, though perhaps not legally, applicable to the redemption of its other forms of currency, has not, since 1882, fallen below the sum of $100,-000,000 until within the last two years.

Secretary of the Treasury Charles Foster, in his annual report[1] on the state of the

---

[1] In this same report (1892) Secretary Foster says further:

" But if $100,000,000 in gold was a suitable or necessary reserve in 1882 and in 1885, it would seem clear that a greater reserve is necessary now. It should be remembered that since 1882 we have added to our silver circulation the sum of $259,016,182 in standard silver dollars coined under the old silver act of 1878. These dollars are nearly all out-

finances of the United States for the fiscal year ending June 30, 1892, says: "In the bank act of 1882 Congress gave expression

---

standing, and largely represented by silver certificates. We have also increased the legal-tender paper circulation by issuing about $120,000,000 of the Treasury notes authorized by the act of July 14, 1890, and to this we are adding about four millions each month in payment of silver bullion purchased.

"It is true that silver certificates are not redeemable in gold, and that the Treasury notes of 1890 are redeemable in coin; but since it has been declared to be the established policy of the United States to maintain the two metals, silver and gold, on a parity with each other, it is obvious that this large addition to our circulation has increased the possible charge upon our gold reserve. In view, therefore, of these increased and increasing liabilities, the reserve in the Treasury for the redemption of the Government obligations should, in my opinion, be increased to the extent of at least 20 per cent of the amount of Treasury notes issued, and to be issued, under the act of July 14, 1890."

From this statement of Secretary Foster, and from the conduct of the Treasury Department since these words were written, it may be assumed that no refusal of redemption of other forms of money in gold would be made where that refusal would have a tendency to create a disparity in the value of the different moneys issued by the Government.

As a matter of fact, the recent large demands for redemption in gold have been based on gold certificates, United States notes, and Treasury notes. Yet, to all practo its belief that $100,000,000 in gold was a suitable reserve by providing that, whenever the amount of gold in the Treasury should

---

tical purposes, the Government is daily redeeming silver certificates and silver dollars in gold, as is admirably explained by Professor J. Lawrence Laughlin of Chicago in the following words:

"It may, perhaps, be asked why the silver dollars of 412½ grains, being worth intrinsically only from 86 to 89 cents, do not depreciate to that value. The Government buys the silver, owns the coin, and holds all that it can not induce the public to receive voluntarily; so that but a part of the total coinage is out of the Treasury. And most of the coins issued are returned for deposit and silver certificates received in return. There being no free coinage, and no greater amount in circulation than satisfies the demand for change, instead of small bills, the dollar-pieces will circulate at their full value, on the principle of subsidiary coin, even though over-valued, and the silver certificates practically go through a process of constant redemption by being received for customs dues equally with gold. When they become too great in quantity to be needed for such purposes, then we may look for the depreciation with good reason."

Professor Laughlin, having written these words in 1884, would probably readily agree that if the policy of the Treasury Department, as stated by Secretary Foster, is maintained, which, among other things, regards silver certificates as a possible charge on the gold reserve, we need never look for any depreciation of silver certificates, notfall below that sum, the issue of gold certificates should cease. In 1885 the then Secretary of the Treasury adopted the practice of reporting $100,000,000 of the gold in the Treasury as a 'reserve for the redemption of United States notes,' and recently the majority of the Judiciary Committee of the present House of Representatives expressed the opinion that, under existing law, the maintenance of this reserve is obligatory." Since Secretary Foster wrote these words, however, the United States Treasury, in following the declared policy of keeping the metals on a parity, has used the gold in the Treasury to redeem not only the "United States Notes" above referred to, but Treasury notes of 1890, which are " payable in coin," and are based on a deposit of uncoined silver bullion. In so doing, the

---

withstanding a possible over-supply of silver dollars and certificates as regards customs payments, or the demand for subsidiary money. We might, however, under such circumstances, expect the necessity of replenishing the Treasury gold supply.

2

Treasury Department has very largely encroached upon what was called the "$100,000,000 gold reserve." Notwithstanding the fact that a $50,000,000 issue of 5 per cent United States bonds was sold by the Secretary of the Treasury within the last year for the purpose of strengthening the gold reserve, that reserve has shrunk, under demands for redemption of other forms of money, to the present sum of about $60,000,000.

There is a strong analogy between the United States and the relation of its currency obligations to its gold reserve and a bank and the relation of its deposits to its cash resources. The Government assumes, in regard to the demands for redemption of other forms of money in gold, what the bank assumes in regard to the demand for the redemption of its deposits in cash — that the actual demand for redemption will never be but a small fraction of the possible demand.

When a bank continues to cash checks over its counter, and thus stands ready to redeem its deposits in cash, then all its deposits are considered good by the community, and checks drawn against them circulate on a par with all other kinds of money. When a bank refuses to cash properly drawn checks over its counter, and thus refuses to redeem its deposits in cash, then the value of all its deposits is lessened in purchasing power, and checks never circulate against them except at a discount, and generally not at all. Thus, as long as the United States Treasurer stands ready to cash Treasury notes, or United States notes, or silver certificates in gold, over his counters on demand, so long will all the money of the United States circulate on a par with gold, and be as good as gold. When, however, the United States refuses this redemption, and by its own act discriminates between its different forms of money, then will the forms of money, other than gold and gold certificates, become discredited and depreciated in value, as compared with gold, and gold money will go out of active circulation.

In carrying out its policy to keep every one of its dollars as good as a gold dollar, by readiness to redeem other forms of currency in gold on demand, this Government is at present holding, on a par with gold, if we include National bank notes, over $1,000,000,000 of paper and silver money which has, therefore, equal purchasing and debt-paying power with gold. It was in pursuance of this policy that Congress repealed the Sherman law under which the "Treasury notes of 1890" were issued, and which was adding about $4,000,000 per month of Treasury notes to the already large amount of other money held at par by the dwindling gold reserve.

We have gone somewhat into detail in speaking of the money issued by the United States Government for reasons which will appear hereafter. Yet this $1,600,000,000 of money does not constitute all of the money which is used by the people of the United States. In fact, it is by far the smallest part of the total money which they use in the daily transaction of their enormous business.

According to the report of the Comptroller of the United States Treasury for 1881, the receipts of the New York banks, on two different days, consisted of 98 7-10 per cent in checks and drafts, and only 1 3-10 per cent of actual cash. The ratio in the United States is given at 94 6-10 per cent of checks and drafts to 5 4-10 per cent of actual money. In a later report of the Comptroller (1890) the proportion in New York of checks, drafts, and bills of exchange to actual cash received was stated, on two different dates, to be 96 4-100 per cent and 95 64-100 per cent, and in the United States at 92 5-10 per cent and 91 4-100 per cent. In his report for 1893 (page 24) the Comptroller speaks of cash transactions as amounting to only 8 7-10 per cent of the total money transactions of the country. Indeed, all authorities agree that, on the average, over 90 per cent of the total business transactions of the citizens of the United States, involving payments of money, are carried on with checks and drafts. The $1,600,000,000 of Government money is actually handled in business to an amount less than 10 per cent of the total amount of money involved in business transactions.

As checks and drafts are always drawn against deposits, the relation of bank deposits to the money used by the people of the country is seen at once to be very important.

In the report of the Comptroller of the Treasury of the United States for 1893, the individual deposits of the State banks, loan and trust companies, savings banks, and private banks of the United States are given as $3,070,462,680, and the individual deposits of the National banks of the United States as $1,465,446,904 (October 3, 1893). In these figures are not included the deposits classed in bank statements under the head, "Due to National Banks" and "Due to State Banks."

Thus we see that the people of the United States, irrespective of whatever actual cash they may have on hand, have a purchasing ability, expressed in deposits in solvent banks and trust companies, amounting to $4,535,909,-584, or a sum nearly three times as great as the total $1,600,000,000 cash which the United States Government has placed in circulation. The greater part of this sum is either subject to check or issued as negotiable demand certificates. While deposits expressed in time certificates and in savings bank balances are not generally devoted to purchases by our people, except under emergency, nevertheless they are regarded as, and under ordinary conditions are, in practice, a cash asset in the hands of any man in relation to their power to pay debts or purchase commodities. If we are to comprehend the monetary situation and money supply in this country, and its relation to the rise and fall in prices, and the prosperity or depression of our business interests, we must fix firmly in our minds this great truth: that since bank-credit currency (to-wit, checks and drafts drawn against bank deposit balances) is accepted on a par with United States currency or coin in payment of a debt, or of the purchase price of a commodity, therefore the supply of money in the country regulating prices can not be considered as United States coin and currency alone, but must be considered as including the total sum of bank credits which, in relation to a commodity or debt, has the same buying and debt-paying power as the actual money of the Government.[2]

---

[2] We must be careful, in considering the effect upon prices of commodities, not to confuse the aggregate of bank deposits and the aggregate circulation, at a given time, of checks and drafts based upon deposits. It is the latter which must be considered. An increase in bank deposits, in itself, would make no difference in prices, if no checks used in the purchase of commodities be drawn against such deposits.

John Stuart Mill says: "Money left with a banker, and not drawn against, or drawn against for other purposes than buying commodities, has no effect on prices, any more than credit which is not used. Credit which *is* used to purchase commodities affects prices in the same manner as money. Money and credit are thus exactly on a par in their effect on prices."

The variation in the total amount of bank clearings in the United States is a very much better index of the varia-

Many of those who assume to discuss financial questions — particularly the silver question — refuse to consider bank-credit currency as related to prices, or to the purchasing or debt-paying power of a dollar, saying "that behind every check there must be the money in the bank." The shallowness of such statements is apparent in the light of the fact that to pay off in cash the deposits of National, State, and savings banks, and trust companies in the United States, there would be required nearly three times the amount of the entire

---

tion in prices than the variation in the aggregate amount of bank deposits. This results from the fact that clearings, to a certain extent, are the measure of the bank-credit money in actual use in carrying on the business of the country, while the aggregate bank deposits are the measure simply of the purchasing power which may be devoted to the transaction of business, but of which a large portion may be absolutely idle, and, therefore, not a factor in the establishment of current prices.

While the aggregate of bank deposits after a panic may soon reach the maximum existing before the panic, it is uniformly the case that a very much longer period of time is necessary to restore aggregate clearings of banks to the old point.

gold, silver, and paper money of the United States Government now in circulation. The business of this country, if done through the same hands as at present in cash alone, without the use of bank-credit money, would demand, in the one city of New York, the presence, every three weeks, of all of the $1,600,000,000 money issued by the United States Government.

How it should happen that the people of the United States have deposited in banks and trust companies nearly three times the entire cash circulation in the United States — in other words, how bank-credit money is created — we will consider in our next chapter. It is in giving to the country the greatest proportion of the money needed to carry on its business that banks render that service which is of the most vital importance.[3]

---

[3] Prof. F. W. Taussig, in the course of a very able treatise upon the "Silver Situation in the United States" (see Vol. 7 (1892), Publications of the American Economic Association, pages 63-64), refers to bank-credit currency in the following language: "The true way to state the con-

ditions on which, in our day, the general range of prices depends, is to compare the quantity of commodities offered for sale with the total volume of *purchasing power* in terms of money. In this volume of purchasing power the largest item consists, in our day, not of actual money, but of credit in various forms.

"In countries like England and the United States it consists chiefly in the form of credit applied by deposit banks — bank deposits and bank checks. That bank deposits and checks are the means of payment in all large transactions, and in many small ones; that the exchanges carried on through them are greater than those carried on with any other form of currency; that they are completed, through the machinery of clearing-houses, with the use of an insignificant amount of coin or notes — these are facts familiar enough. Indeed, the effects of credit as a substitute for money may have been explained in economic text-books so fully that they may be assumed to be well understood; it is the extent rather than the nature of the effects that needs to be insisted upon. At any one time, and for considerable lengths of time, the general range of prices, with a given volume of transactions, depends not on the quantity of money simply, but on the volume of credit used as purchasing power; and of this volume of credit the chief item in countries like England and the United States is bank currency in the form of deposits and checks. In such a state of things the increase of other forces of currency can have, in itself, only a minor effect."

W. Stanley Jevons, in his work entitled "Money and the Mechanism of Exchange," quotes from Prof. Cliffe Leslie's article in the *Fortnightly Review* for November, 1870, as follows: "The improvements in locomotion and in commercial activity, which have so largely augmented the

money-making power of the Germans, have also quickened
prodigiously the circulation of money; and the develop-
ment of credit, likewise following industrial progress, has
added to the volume of the circulating medium a mass of
substitutes for money, which move with greater velocity.
A much smaller amount of money than formerly now suf-
fices to do a given amount of business or to raise prices to a
given range; and to the increased amount of actual money
now current in Germany we must add a brisk circulation
of instruments of credit.   Were the circulating medium
composed of coin alone, whatever the amount of the
precious metals issuing from the mines, or circulating in
other countries, and whatever the price of German commodities in market abroad, no rise in the prices of German
commodities at home could take place without additional
coin to sustain it."

John Stuart Mill, in his "Principles of Political Economy," says: "In a state of commerce in which much credit
is habitually given, general prices at any moment depend
much more upon the state of credit than upon the quantity
of money.   For credit, though it is not productive power,
is purchasing power; and a person who, having credit,
avails himself of it in the purchase of goods, creates just as
much demand for the goods, and tends quite as much to
raise their price, as if he made an equal amount of purchases with ready money."

CHAPTER III.

HOW BANK-CREDIT MONEY IS CREATED.

Every National bank in the country is presumed, by law, to be on a conservative and safe basis, so far as its ability to redeem deposits is concerned, when it has on hand in cash, and on deposit with approved banks, called reserve agents, 25 per cent of the amount of its total deposits.

In cities of small population, a less proportion of cash and cash resources to total deposits is legally deemed consistent with conservative banking. When the cash and reserve accounts (called together "cash resources") exceed to any extent the 25 per cent limit, the bank, according to the generally accepted rule, can safely loan the excess. In the determination of its ability to loan,

a bank does not ordinarily consider the relation of the amount of cash in the vaults to
the amount of its credits with reserve agents,
for the reason that in usual practice the
reserve agent will ship cash when called for,
or the bank can sell for cash, to neighboring
banks, exchange drawn against its balance
with the reserve agent.   It is the relation
of the sum of the cash and reserve balances
to deposits which determines whether or not
the bank is in a condition to loan.   While
this is a well-known fact, the consequences
of it are often overlooked.

The fact that National and other banks
are allowed to loan on the strength of their
deposits in other banks, as well as on account
of cash actually in their vaults, means that,
considering banks as a whole, the actual cash
held by them, into which checks and drafts
are by their terms and by practice concurrently convertible, is much less than 25
per cent of the deposits.   This fact, together
with the fact that one bank will credit,

as a deposit redeemable in cash, a check on another bank, means also that our banks add to the purchasing power of the community, or, in other words, create money for the community. In giving the following illustrations of those operations of banks which result in an increase of purchasing power to the community, we do not wish our readers to infer that, under similar circumstances, all banks would act as do the banks which we assume as existing for purposes of illustration. Yet whether it be the policy of a bank to run with a 25 per cent reserve, as do the banks in our illustrations, or a 40 per cent reserve, the general effects of loaning operations are the same.

The instances we will give clearly demonstrate the enormous influence of banks upon the money supply or purchasing power of a community.

We will suppose a National bank to be organized at Chicago with $600,000 capital, all of which, at the commencement of business, is in the vaults in cash. Let us assume that the first action of the Chicago bank is to deposit $300,000 of its capital in the New York bank which has been approved by the Comptroller of the Treasury of the United States as its reserve agent, the latter bank having already on hand 25 per cent of its deposits, in cash and cash resources, at the time the deposit of $300,000 is made by the Chicago bank.

The statement of the Chicago bank then stands as follows:

ASSETS.		LIABILITIES.	
Due from approved reserve agents	$300,000	Capital stock	$600,000
Cash in vaults	300,000		
	$600,000		$600,000

We will assume that during the next week the Chicago bank receives $800,000 in deposits and makes $1,000,000 of loans, retaining 50 per cent of its deposits in cash resources. The statement of the Chicago bank then stands as follows:

ASSETS.		LIABILITIES.	
Loans and discounts..$ 1,000,000		Capital stock..............$ 600,000	
Due from approved reserve agents in New York	300,000	Deposits...................	800,000
Cash in vaults.........	100,000		
	$1,400,000		$1,400,000

We will assume that of the $300,000 deposit made by the Chicago bank, the New York bank loans $225,000, reserving $75,000 in order to still keep its cash reserve 25 per cent of its deposits. We can now trace the increase in the purchasing power of the country involved in these transactions, or, in other words, ascertain the increase in bank-credit money.

It is as follows:

1st. An increase in New York of $225,000 purchasing power, loaned by the New York bank to borrowers there, since but $75,000 of the $300,000 deposited by the Chicago bank is retained by the New York bank to protect itself on the Chicago deposit, and since at Chicago this $300,000 New York balance does

3

the work of an equal amount of cash in assisting to hold good, under ordinary business conditions, the $800,000 deposits of the Chicago bank.

2d. An increase of $400,000 purchasing power, or bank-credit money at Chicago. This increase has been created in this way: In making $1,000,000 loans, the Chicago bank, in addition to loaning to its customers its capital of $600,000, thus returning to the community all the purchasing power it possessed before the organization of the bank, has also loaned to them $400,000 of the $800,000 placed in the bank by depositors, thus adding $400,000 to the purchasing power of the borrowers. And yet, by this latter transaction, no purchasing power has been taken away from the depositors, since checks drawn against their $800,000 deposits are accepted in payments on a par with the gold, silver, and paper money of the Government, the bank having pledged itself to redeem them in cash on demand. Thus we see that this Chicago bank, so conservative as to hold in cash and reserve agents one-half its total deposits, has yet, by these simple transactions, increased the supply of money which can be devoted to the purchase of commodities, or the payment of debts, in the sum of $400,000 at Chicago and $225,000 at New York — a total of $625,000, being a sum $25,000 in excess of its original capital of $600,000.

Can there be a better illustration of the wonderful influence of the banks of the country on business and prices?

A simple example of the steps by which banks increase the purchasing power of a community is the following: A customer of this Chicago bank borrows $60,000, which, being 10 per cent of the capital stock, is the maximum amount which can be legally loaned to one individual, giving his note with collaterals, and receiving a credit upon his bank book of $60,000, less the interest discount, as is the general custom. He does not draw the actual cash from the bank, but gives checks against his balance, and these checks are re-deposited by others, the check and clearing-house system making the use of the actual cash unnecessary as between the first borrower and his creditors, and between his creditors and those with whom they settle. Leaving out of account the interest discount paid, the bank, by this transaction, increases its deposits $60,000, and increases its loans and discounts $60,000 — its assets and liabilities being thus increased in the same amount. It increases the purchasing power of the community by the sum of $60,000, but it has reduced its loanable funds by $15,000 (to-wit, 25 per cent of $60,000), which must be held in reserve, either in cash or in other banks, as a protection against the possible calls for cash upon the additional deposit of $60,000. The net increase of purchasing power in the community by this transaction is therefore $45,000. It has not found it necessary, in this transaction, to draw any portion of its balance in New York, or to pay out any cash.

This creation of money (for such is the true character of the transaction) is based upon two assumptions: 1st. That the borrower is able to turn his note into cash at maturity. 2d. That the bank is able to turn the checks into cash upon demand.

We have already shown that bank deposits, since they are the basis of checks and drafts, have a great deal to do with the money of the country. We now see that bank loans, since they form the basis of a large part of bank deposits, play a great part in determining the money supply of our country.

Having considered the connection between the increase of loans and the increase in deposits, and the consequent increase in the purchasing power of the community, let us now consider the connection between a decrease in deposits, and the resulting decrease in loans, and decrease in the purchasing power in the community. Let us take the case of a New York bank which is called upon to ship $150,-000 of currency to one of the banks of Chicago depositing with it. Before the shipment we will suppose the following to be a condensed statement of the condition of the New York bank:

ASSETS.		LIABILITIES.	
Loans and discounts...	$1,000,000	Capital stock...........	$ 250,000
Cash resources.........	250,000	Deposits...............	1,000,000
	$1,250,000		$1,250,000

At the time of paying the Chicago depositor $150,000, the cashier calls in $150,000 of loans, which, for the sake of our illustration, we will suppose are paid by checks on his own bank. In calling in $150,000 loans, the bank takes that much bank-credit money, or purchasing power, from the community. Its loans and discounts are reduced $150,000, its cash resources $150,000, and its deposits $300,000, leaving the statement of the bank as follows:

ASSETS.		LIABILITIES.	
Loans and discounts	$850,000	Capital stock	$250,000
Cash resources..........	100,000	Deposits	700,000
	$950,000		$950,000

But we readily see that the bank must contract its loans still more, since the remaining

---

[1] See note at end of chapter, page 42.

$100,000 cash resources lacks $75,000 of being 25 per cent of its remaining $700,000 deposits, to-wit, $175,000.

Therefore the bank must still further contract the purchasing power of the community by creating again the proper ratio between cash resources and deposits. Assuming again that the borrowers pay the loans which are called, by checks on the same bank, it is evident that in this case, to restore the 25 per cent proportion of cash resources to deposits, loans must still be contracted $300,000, decreasing deposits $300,000, and taking that amount of purchasing power away from the borrowers. Whether the loans are paid by checks on the bank calling them, or by checks on other banks, the effect on the purchasing power of the community is the same. The statement then stands:

ASSETS.		LIABILITIES.	
Loans and discounts	$550,000	Capital stock	$250,000
Cash resources	100,000	Deposits	400,000
	$650,000		$650,000

By these transactions, therefore, the purchasing power in New York is diminished by $450.000. This amount exactly equals the possible increase in purchasing power given to Chicago, if the Chicago bank chooses to use the $150,000 cash to its limit, as a basis for creating bank-credit currency. It can loan credits on its books (deposits) to an amount of which the $150,000 cash is 25 per cent, to-wit, $600,000. As the $150,000 deposit in New York, before it was drawn by the Chicago bank, was counted a cash asset, the possible net gain in Chicago really consists of the difference between $600,000 and $150,000, or $450,000.

It is readily seen from this examination into the method of the creation of bank-credit money, which permits the purchasing power created by loans, and expressed in a draft or check on one bank to be deposited as a cash asset in another, that deposits in banks eventually greatly exceed the total cash supply of the country which is furnished by the Government.

We pointed out in a former chapter the extent
to which deposits have outgrown the cash
supply of the United States. The reason why
the exact nature of the effects resulting from
the ordinary transactions of banking, similar
to the ones forming the basis of our illustrations, is not better understood, is because, as a
general rule, the increase and reduction of
loans and deposits and bank-credit currency
take place quietly, and in obedience to the
normal demands of business communities,
according as they need a greater or less amount
of money in the transaction of their legitimate and ordinary business. The transfer of
currency and credits between banks and different localities under normal conditions is
largely determined by the fluctuations in interest rates in different localities, and in the common course of business these transfers offset
each other, and all goes well. Not until times
of panic come, when the banks of our country,
in keeping faith with the depositors, whom
they have promised to pay "on demand," are

compelled not only to practically cease distrib-
uting purchasing and debt-paying power in the
shape of loans, but to destroy it by calling in
loans already outstanding, do we see the almost
unlimited dependence of our people upon our
banks, and the true value of banks as regula-
tors of the business machinery of the country.

---

[1] New York and Chicago are "Central Reserve" cities,
and their National Banks are required by law to keep 25
per cent of their deposits in *cash in their vaults*. The
illustration, therefore, would technically be more accurate
if the banks named were assumed to be in Omaha and New
York instead of Chicago and New York.

CHAPTER IV.

In these days, when all exchanges are made in terms of money, there is of course no other one thing in the business world in which people are more interested than in the prices of commodities, measured in the terms of the money of the land. As prices rise and fall, so do the hopes and wealth of the business man. In the relation of the price of a certain commodity, which must be paid the wholesaler, to that price which can be demanded from the customer, lie the profits of the great multitude of tradesmen. In the relation of the prices of raw material to the prices which the manufactured article will demand from the public, depend the profits of the manufacturer and the wages of the laborer. The relation of the price of to-day to the price of to-morrow, and its effect upon the relations of debtor and creditor, is always one of the continuing problems of a commercial age. The laws and conditions which govern the fluctuations in prices are always those which engage the earnest attention of business men. Were the supply of money inflexible in amount and uniform in value, the price of an article would depend simply upon the relation of the demand for the article to its supply. But it is evident that in addition to the demand for, and supply of, an article, its price must depend also upon the demand for, and supply of, money, in terms of which the price must be expressed.

This being true, the fluctuations in the money supply of a country have as direct an effect on prices as the fluctuations in the demand for, and supply of, an article. To determine, therefore, the true effect which monetary conditions are having upon prices of commodities, we must look to the fluctuations in the money supply, and to the cause of the fluctuations. And we must remember, also, that the supply of money regulating prices must, as we stated in a former chapter, be considered as including the total sum of all bank credits which, in relation to a commodity, have the same buying power as money.

When we admit this we see immediately that the fluctuation in prices, caused by fluctuations in money supply, is largely determined by the nature of the fluctuations in the amount of bank-credit currency in circulation. From the reasoning employed in our preceding chapter it is easily seen that the banks have such power to expand the circulation of bank-credit money by making loans in obedience to demands for money that, under ordinary conditions, an increase in the demand for money will be met with a corresponding increase of bank-credit money— the tendency thus being to keep the purchasing or debt-paying value of a dollar uniform. Accordingly when business does not need such an amount of money, the community pays off its loans to save interest, and there occurs the reduction, both in loans and deposits, which we have before shown means a decrease in the total amount of purchasing power of the people. We can see, therefore, that under normal conditions the increase or decrease in bank-credit currency has a tendency to keep the purchasing power of a dollar stable, since these increases or decreases in money supply correspond with the increases or decreases in the demand for money. Owing to the fact, however, that in times of easy money, loans are unfortunately made in obedience to speculative demands, as well as in accordance with the legitimate demands of business, prices during a period of expanding loans have a tendency to rise higher than they would if loans were confined to the supplying of legitimate business alone. In such times, prices have a tendency to become fictitious. During a period of contracting loans, prices have a tendency to remain

stationary, or to fall. Under a forced contraction of loans, prices fall with extreme
rapidity. There is a remarkable correspond-
ence between the rise and fall in prices and
the increase and decrease in the bank loans
of the United States, which latter, of course,
is the best index of the increase or decrease
in the bank-credit money of the country.
During the year 1893 we had a remarkable
illustration of the extreme sensitiveness of
prices to a contraction of loans and a decline
in bank deposits.

According to the report of the Comptroller
of the Currency for 1893, the deposits in the
National banks of the country, between May
4 and October 4, 1893, were reduced to the
enormous extent of $378,119,563. The con-
traction in loans, made necessary by the withdrawal of these deposits, was $318,767,691 ; the
contraction in balances on deposit with other
banks was $51,198,856; the contraction in
stocks and securities was $2,177,912. The
banks also took out $31,265,616 of new circula-

tion, and borrowed in various ways $36,615,092. All this contraction, added circulation, and borrowing was made necessary by the withdrawal of National bank deposits alone. The enormous loss in purchasing power of the community, caused by the panic of 1893, is difficult to estimate, owing to the fact that it is impossible to ascertain from the very complete statistics regarding National banks collected by the Comptroller, or in any other way, the proportion of deposits which were paid out in money and hoarded, and the proportion of the decrease in deposits which resulted in the canceling of deposits against loans — an operation which, while it lessens the possible demands upon a bank, actually nets the bank no cash, and destroys in large proportions the purchasing power or bank-credit money of the community in the manner shown in the preceding chapter. This much is certain, at least, that the contraction of National bank loans, in this period, of $318,767,691, represented a net loss of that much purchasing power in the United

States. To this loss must be added another enormous loss, resulting from the decrease in private and savings banks deposits and loans. In his report for 1893, the Comptroller says: "The fright among depositors of the present year appears to have affected all classes of banking institutions alike. The shrinkage of deposits of National banks, from May 4th to July 12th last, exceeded $190,000,000. In order to ascertain the extent of the shrinkage in banks other than National, the Comptroller requested each State officer, charged with the supervision of banks organized under State authority, to submit a statement showing similar information with respect to that class of banks. Replies were received from the officers of twenty-three States and two Territories, indicating that the loss to banks of this character corresponded with that shown by the returns from National banks. Generally no information was given with respect to savings banks, and much less regarding private banks."

4

Remembering that the deposits in other banks, here referred to by the Comptroller, are about double the deposits of the National banks of the country, we feel justified in assuming, after making due allowance for the fact that the deposits of banks other than National are not subject to check to such an extent as are those of National banks, that the loss in purchasing power, or bank-credit currency, due to the withdrawal of deposits in banks other than the National, was at least $350,-000,000, making the total reduction of bank-credit money, due to the panic of 1893, aggregate at least the sum of $668,000,000. We feel that this sum falls far short of expressing the actual loss in bank-credit money, or purchasing power, caused by the panic of 1893. The low range to which the prices of commodities in the United States were reduced, during the year 1893, was caused by this great decrease in bank-credit currency. The great extent to which they were reduced, and the rapidity with which they fell, is well known.

CHAPTER V.

THE RELATION OF BANK-CREDIT CURRENCY TO THE SILVER QUESTION.

The great desideratum, in the matter of money, is to obtain a money whose purchasing and debt-paying power remains as stable as possible. As stated before, the value of money is determined by the same law of supply and demand which operates in the case of ordinary commodities. While such a thing as a dollar of absolutely uniform purchasing and debt-paying power is unattainable, it is in the power of Government to prevent a very great change or fluctuation in the value of a dollar by wise and proper laws. Perhaps the questions of greater importance involved in the present discussion of the silver question are as to the effect of free-silver coinage law

upon general business conditions, and upon a
contract for a time payment of money, or, in
other words, upon the debtor and creditor
classes. The effect upon the laborer and the
price of labor, caused by an enhancement or
depreciation in the value of a dollar used
simply as a medium of exchange, is not of so
great importance, since the prices of labor and
the prices of commodities, into which labor is
exchangeable through money, are affected
equally and alike by the change in the standard of measure. If the price of labor, through
an enhancement in money, falls, the price of
other commodities falls. If the price of labor
rises through a depreciation in the standard,
the price of other commodities rises also.[1]

---

[1] Hon. J. K. Upton, in an able work entitled "Money
in Politics," pages 144-145, has the following remarks and
table: "To those who believe that the aggregate exchange
value of a circulating medium can be increased or diminished by the will of Congress, or any human agency, is
commended the fact that after sixteen years of legisla-
tive effort, by which the face value of the circulation was
reduced more than $300,000,000, the coin value was reduced

Therefore the chief evil, apart from the injury done general business, which results from a disturbance in the value of the standard as measured in commodities, is found in its disturbance of the relation between the debtor

---

less than $6,000,000. The great law of demand and supply fixed the amount needed and mocked the futile efforts of those who tried to overrule it. The following statement shows in tabular form the changes which took place in the amount and valuation of the paper circulation for the years named:

STATEMENT SHOWING THE AMOUNT IN MILLIONS OF OUTSTANDING PAPER CIRCULATION, AND ITS VALUE IN COIN, TOGETHER WITH THE VALUE IN COIN OF ONE DOLLAR IN PAPER, AT THE CLOSE OF EACH FISCAL YEAR, FROM 1865 TO 1879, INCLUSIVE.

Year Ending June 30.	Amount of Circulation. Millions.	Coin Value of Circulation. Millions.	Coin Value of One Dollar of Paper.
1865	983	697	.71
1866	892	589	.66
1867	827	587	.71
1868	770	540	.70
1869	756	552	.73
1870	745	633	.85
1871	748	665	.89
1872	738	646	.87½
1873	750	648	.86¼
1874	781	711	.91
1875	773	674	.87
1876	738	656	89
1877	698	662	.95
1878	688	684	.99½
1879, January 1st	686	686	1 00

and creditor, between whom there is a contract calling for a second transfer of money after a certain time. If money appreciates in value, it takes more goods to get the money to pay the debt with than the money borrowed could buy, and the debtor is wronged. If money depreciates, it takes less goods, and the creditor is wronged. The question, however, of the increase or decrease of the amount of the standard money — gold — does not play such an important part in burdening or relieving the debtor class, as many of the advocates of free silver would have us believe.

A characteristic of free-silver arguments, which seems to be common, is an apparent assumption that the purchasing power of a dollar in a gold-standard country, whether that dollar be expressed in a legal-tender note, a silver coin, or a bank draft, is wholly determined by the value of gold, without the value of the gold being affected by the concurrent use of the silver and paper money, which does the work of gold money. Tables

are given to show the ratio of gold increase
to population increase, the amount of gold
increase to increase in property values and
business transactions, the implication being
that the failure to pass a free-silver coinage law throws upon gold a burden which
has been growing faster in proportion than
the supply of gold has been increasing,
thereby adding enormously to the value of
gold, and therefore to the value of money,
thus working an injustice to the debtor
class. Until this argument is analyzed, it
seems strong.

Its chief error consists in this: That it fails
to take into consideration bank-credit money,
which is the most influential kind of money
in determining the ease or difficulty with
which debts are paid. This kind of an argument seems to be based upon a false premise.
That false premise is that a gold standard is
something that stands off by itself, as it were,
and that any increase or decrease in its
amount increases or decreases the value of

every dollar in circulation in the proportion of the amount of increase or decrease of gold to the total amount of gold.

Always assuming that the Government of the United States maintains in its Treasury, as under its present policy, the comparatively small amount of gold necessary to meet any demands for redemption of its other currency, which, in the light of experience and existing conditions, may be made, the fact is that, under our money system, the increase in the world's gold supply, or our own gold supply, cuts very little figure in the determination of the purchasing or debt-paying value of a dollar.

No system of logic can disprove the proposition that the value of money depends upon the law of supply and demand. If, therefore, the supply of money is increased by bank-credit currency, the value of money measured in commodities, including gold money, is decreased; and correspondingly, when bank-credit currency is contracted, and the total supply of money is thus diminished, the value of all money, including gold money, is increased.

When we remember that the proportion of bank-credit money used in business is ordinarily over ten times the amount of cash money, we can see that the fluctuations in bank-credit money have a greater effect upon the purchasing and debt-paying power of actual money than any other one cause. The effect of the annual increase or decrease in the output of gold at the mines, or as money, pales into insignificance in comparison. Thus we see that where the banking business is extended, as it is in this country, and where the Government, for the purpose of keeping ample the money supply, injects its credit, with proper restrictions, into the currency, as in this country, the fact that we use a gold standard, and are on a gold basis, and repealed the Sherman law, which was becoming a menace to our ability to keep on a gold basis, does not justify any one in assuming that a crime against the debtor class is being committed.

The more the money question is studied, the more does the fact become apparent that, where such an immense volume of business is transacted through the banks, the most important factor which goes to make up the purchasing and debt-paying power of that most uncertain measure, the dollar, is the condition of the banks, and the relation of public confidence to the banks, upon which the banking business is based. Common experience teaches us this truth. When does money become more valuable, and goods, measured in money, less valuable, than in a bank panic, when depositors are "running on the banks"? When deposits, which form the largest basis for bank-credit money, are withdrawn, the value of a dollar, under the operation of the law of supply and demand, increases, and the prices of wheat, corn, pork, iron, and other commodities go down almost with a rush.

Advocates of the free coinage of silver give tables covering many years, by which they attempt to show, by a decrease in cost of wheat

and corn, etc., an increase in the value of a
dollar in a gold-standard country, or, as they
unjustly term it, an increase in the value of
" gold," using it as synonymous with money.

The impossibility of gauging accurately the
value of a commodity like gold, which is subject to the law of supply and demand, by its
average equivalent in other commodities also
subject to their laws of supply and demand,
especially when the value of gold money is
influenced by the supply and demand of other
money doing the work concurrently with gold
and in place of gold, the volume of which
other money fluctuates constantly, has always
seemed to us very manifest. And when we
contemplate the immense fluctuation which
has occurred in the purchasing and debt-paying power of a dollar, during this last year, in
this " gold standard " country, we are still more
inclined to doubt the reliability of these tables,
whose makers have assumed such a scientific
air.

The fact of the matter is that, when public

confidence in the banking institutions of this country exists, the bank-credit currency, being flexible, is, as one might almost say, automatically regulated in volume according to the needs of business ; and this fact has much more of an effect in keeping stable the value of a dollar than the arbitrary amount of increase or decrease in the annual output of either gold or silver, in their relation to population or property increase, can ever have. When panic conditions are absent, and when money is being loaned by our banks in amounts necessary to supply the needs of business, we seldom hear any complaint about falling prices and the " robbery of the debtor."

When, after a panic, confidence in banking institutions has been restored, money commences to flow out from the banks in loans, and bank-credit currency slowly expands to meet the demands of a recuperating and growing business. A panic, such as we have just passed through in 1893, is, as has often been said, the result of overtrading. Credits which, with the aid of banks and the naturally venturesome disposition of the ordinary man, always increase and grow faster than governments ever issue money, eventually become so out of proportion to values, and the amount of money in circulation, that their scaling down can not be accomplished without creating widespread distrust, which attacks the worthy along with the unworthy, and withdraws from business, money, just at the time it is most needed in the re-adjustment of too expanded credits. It is considered an able argument by some to point out the immense expansion of credits — both governmental, railroad, and private — reaching into the tens of billions of dollars, and then to point out the disproportion of debts, to be paid in money, to the amount of money which is available to pay them, especially of gold, using this as an argument to show the necessity of more money. Unfortunately, the more money governments issue or coin, the greater becomes the proportionate growth of credits. History conclusively shows

that an inflation of the currency, or, in other
words, any large amount of money in circulation in excess of the ordinary amount, is fol-
lowed by what is called an era of speculation,
in which great bubbles of credit are blown, the
bursting of which inevitably brings a panic
with all its dire results. In other words, if,
with the idea of lessening the disproportion
between credits and money in times of general
business prosperity, you increase the amount
of money in circulation, instead of using the
increased amount of money in the liquidation
of existing indebtedness, society uses it as a
basis for the creation of additional credits,
bringing about a worse state of affairs than
before. The only time when a sudden increase
of money in large sums is beneficial to a community is when the process of liquidation has
commenced, when the bubble is burst, and
when an increase of money operates to build
up a stronger foundation for the castle of business rather than to build fancy turrets on its
top. When we leave out of account the credit

substitutes for money, so enormous is the amount of governmental, railroad, and private debts of the world that, for all practical purposes, the sum of all the gold and silver in the world is as inadequate to pay them as either gold or silver alone.

The truth of the matter is that no man lives, no government exists, which can devise an absolutely equitable currency system. No inflexible system can be ample enough to accommodate itself successfully to what are known as panic times, and not lead to over-speculation in other times. The crying need of the day is for a currency so flexible in the hands of the Government that, when bank-credit currency is diminished by widespread withdrawal of deposits, and life-blood is being drawn from business, it can be increased by the Government under some such plan as exists in England to-day, or under some other proper restrictions, and in some way used to lessen the awful process of such a liquidation as that of 1873 and 1893.

The need of the existence of such a power in our Government is shown by the enforced creation of a substitute for money, in the shape of clearing-house certificates, by the New York and Boston clearing-houses during the last summer. That this is the question of the day, and not the simpler one of a standard, is easily shown. Populists and others, seeing the evil effects of our too inflexible currency, have assumed these evil effects have largely arisen from the demonetization of silver, and the lack of that contradiction in ideas, a "double standard."

To convince us that the increase of money, incident to the passage of a free-silver coinage law, will not correct all the evils to which the debtor is subject under our present system, we have but to remember that in the short period of five months of the year 1893 there was destroyed, by the panic in the United States, bank-credit money to the amount of at least $668,000,000, a sum equal to the total silver production, at the present rate, of all the silver mines in the United States for over nine years, not sparing any for use in the arts.

An intelligent debtor soon comes to understand that those influences and forces which tend to unsettle general confidence in monetary conditions, are the causes which lead to his greatest oppression and loss. Such causes tend not only to stimulate creditors (including bank depositors) into calling for their principal, but to frighten others from loaning and investing, thus making re-borrowing impossible, and necessitating often the sale of securities and property at a great sacrifice, since they must be sold when the public are afraid to buy.

Of all influences which are injurious to debtors, those which tend to bring about a change of monetary standards are the worst, especially in the United States, where such a large amount of silver and gold is circulating side by side as at present. Under the operation of Gresham's law, which is that,

5

under such conditions, the cheaper money will drive the dearer money out of circulation, there would be driven out of our circulation, by a lapse on the part of our Government from the present gold standard to the cheaper silver standard, our entire gold circulation, amounting to about $500,000,000.

A change of standards unsettles credits; it unsettles banking; it unsettles business; it unsettles international exchanges unless all nations act as a unit in the change. We are unable to see how a change of standards, in the United States alone, can prove of either temporary or eventual benefit to the debtor.

CHAPTER VI.

ONE PRESENT NEED OF OUR NATIONAL BANK-
ING SYSTEM.

It has been our effort to point out in this
little volume the extent of the dependence
of the prosperity of our nation and our people
upon the normal operation of our present
banking system. The most disastrous blow
which the business interests of our country
can receive, consists in the weakening of the
confidence of the depositors in the ability of
our banks to redeem deposits.

Under our present laws and business
methods, banking and industrial panics have
occurred together. If it were not for the
necessity upon the banks of paying frightened depositors during an industrial panic,
the solvent and worthy institutions of the

country would never be forced into liquidation because of an inability to turn unquestioned collaterals into cash. Under our system of doing business, times of liquidation must periodically occur when too expanded credits must be adjusted to true values. It should never be admitted, however, that these periods of general liquidation must be accompanied, as they have generally been heretofore in this country, by runs on the banks and the withdrawal of active capital from business to go into old stockings or safety-deposit boxes, at a time when it is most needed in the general re-adjustment of credits. Could banks, in times of business liquidation, be assured that they would not have unusual demands for deposits, they could successfully maintain thoroughly solvent individuals or corporations from financial destruction, which sometimes comes to them from demands for quick payments from frightened creditors. To so build up the confidence of the people in the safety of National banks as depositories that, in times of business panic, deposits would not be withdrawn — to make such a thing as a bank panic impossible — is, perhaps, something which is difficult of accomplishment.

There is one law, however, which has already been considered somewhat in Congress, the passage of which, in a proper form, seems to us of vital importance. That law is one which should levy a tax upon National banks for the purpose of creating a fund to be held by the United States Treasury for the re-imbursement of depositors in failed National banks. National banking statistics show that a fund of the necessary amount would soon be created by a comparatively small tax upon each National bank. The Government, in the interests of the fund, would, of course, take an assignment of the claims of depositors against the assets of the failed institution. The levy of the annual tax on each bank for the creation of such a fund should most equitably be based upon the average line of deposits of the bank for the preceding year. This system would seem to be more equitable than to base the tax upon the capital of the bank, or in any other arbitrary way. It must certainly be admitted that the establishment of such a fund would have a tendency to prevent the mad rushes of small and large depositors during times of panic for money, which they hoard away in safety-deposit boxes or other hiding places. A certain loss of deposits will always necessarily be incident to industrial panics; but we will all agree that the deposit losses which are most acutely felt by the business community consist of those which are paid out over the counter in cash to panic-stricken customers, who do not draw their deposits because of other demands upon them, but because they fear the loss of their money if they allow it to remain.

Such a law might contain provisions for a more rigid system of National bank examination, and still further contract the powers of the active officials in their rights of borrowing from the bank under their management. Such a law would probably have the effect of increasing the deposits in National banks, and thus offer to them compensations for the tax levied upon them for the creation of the fund.

If the effect of such a law would be to render bank deposits more stable under all conditions, as in our judgment would be the case, no law could be of more value and importance to the debtor, or to the creditor, or to the community at large. If such a law had been in force for several years prior to and during 1893, there might have been now in prosperous existence, hundreds of the then solvent business institutions and individuals ruined by the panic of that year. Besides the great importance of the law, as related to the general prosperity of the country, it is meritorious, as preventing the keen suffering in those localities where bank failures occur, and where the hard-earned savings of the community, under our present laws, are often swept entirely away. By such a law, losses are distributed as by insurance, the beneficial effects of which need no argument.

To the passage of such a law in proper form, we trust the efforts of Congress will be directed.

APPENDIX.

In submitting the preceding pages to a friend for criticism, the writer received such a concise and able statement of objections to his views that he gives it here, together with his answer.

The criticism, which I have divided for purpose of reference into two parts, reads as follows:

PART A.—"The propositions open for discussion would, of course, be those which relate to the relative power, in determining prices, of money, and the various forms of credits which, by the operations of banking, are made to take the place of currency or circulation. Granting that the amount of these is very much larger than the amount of real money, do they become, therefore, potent in the ratio of these amounts in controlling and determining prices? But prices are the measure of the value of money, meaning neither

more nor less than the scale of exchange of money
for commodities.　The question, then, may take
the form:　Does the gross amount of bank and
other credits fix the value of *money?*　We must
be careful to distinguish here between intrinsic
value and the rate of interest — the former is
what we are now dealing with.

"Your argument seems to be this: If the forms
of credit used in trade are a hundred to one greater
than the real money in circulation, the intrinsic
value of the money is of almost no importance,
for the influence of the circulating credits is a
hundred times more, upon prices.

PART B.—"Let us test this:　Suppose a government, not having the historical lessons about
debasing the coin before its eyes, should so de-
base its gold and silver money as to make it practically a pewter coinage of the same size, etc.,
what would be the result?　Would you say that, as
the coin is only in nominal amount one-hundredth
as much as the forms of circulating credits, therefore the debasing is of little comparative importance?　What is each note or bill of exchange but

a promise to pay the legal money? *Its* value (the note or bill) is only so many of the standard units. If those units have been diminished in value, *all* your notes and bills have suffered the same deterioration. If the coin has been reduced to one-hundredth of its value, the gross amount of all outstanding forms of credit has been reduced in like ratio. You may have cut down a million in coin to ten thousand. Can you say that the loss is only 990,000 out of a gross circulation of 101,000,-000? No; you have, by the same act, taken the 99 per cent out of the value of every note and bill of the whole hundred millions.

"As a question of exchange, it seems to me to be essentially the same as if your notes and bills were payable in another commodity, say wheat. If wheat has gone down from $1 to 50 cents a bushel during the life of your paper, the makers will only pay half as much to take it up. It is this relation of all *promises to pay* to the *value* of the *unit* in which *payment is to be made* which seems to me to make it fatally erroneous to treat these promises to pay as if 'real money.' In every 'corner' of a market the tremendous nature

of the transaction and its consequences is determined by the price of the *article*, though there
may be very few bushels actually in the elevators,
etc. This seems to be the heart and soul of the
whole theory of a standard. If, by any means,
natural or artificial, the value of the *standard* is
changed, you affect, in the same ratio, the value of
all outstanding contracts to pay in that standard."

ANSWER:

1. To that part of criticism marked "A"—

You state that "Granting that the amount of
these (*i. e.* banking credits) is very much larger
than the amount of real money, do they become,
therefore, potent in the ratio of these amounts in
controlling and determining prices?" Farther
on: "We must be careful to distinguish here
between intrinsic value and the rate of interest—
the former is what we are now dealing with."

Here, in effect, I think, you ask a distinction to
be made, for the sake of argument, that is not
made in the transaction of business by the community. If, in purchasing a commodity, I pay

with an instrument of credit, such as a bill of exchange, which does the work in the transaction of an equal amount of the standard money — gold — from an economic standpoint the resulting effect upon prices, caused by the transaction, is the same as if gold money had been used. The fact that the business community accepts indifferently either gold money, which is intrinsically valuable, or a bank draft, which is not intrinsically valuable, and makes no distinction between them, makes, in the resultant effect upon prices, the drafts more of a factor than the standard money. And any increase in the money supply or purchasing power in use, if it may be so expressed, whether it be credit money in the form of checks, or gold money, has its direct effect in fixing the value of money as measured in commodities. The principle involved in the relation of the various forms of credit to prices for which we contend is well established. John Stuart Mill, in his "Principles of Political Economy," states it in the following strong language: "In a state of commerce in which much credit is habitually given, general prices at any moment depend much more upon the state of credit than upon the quantity of money. For credit, though it is not productive power, is purchasing power; and a person who, having credit, avails himself of it in the purchase of goods, creates just as much demand for the goods, and tends quite as much to raise their price as if he made an equal amount of purchase with ready money."

Now, taking your own statement —"but prices are the measure of the value of money, meaning neither more nor less than the scale of exchange of money for commodities"— in connection with the well-recognized principle here stated by Mill, I answer your question —"Does the gross amount of bank and other credits fix the value of money?"— as follows:

The supply of money, regulating, in accordance with the demand, its value, includes bank-credit money, as defined before, as well as actual money (to-wit, money issued by the Government). Therefore the value of all money, expressed in commodities, is fixed more by the gross amount of bank credits in commercial use as money (since they constitute the bulk of our money) than it is by the amount of standard money in use, provided always that the par of exchange is theoretically and practically maintained, as under present conditions. To make a distinction in the effect upon prices, and therefore in the effect upon the value of money, between increases or decreases in the supply of actual money, and in the supply of credit substitutes for money, is impossible. The sum of the two fixes the value of money as measured in commodities. An increase or decrease in the supply of either affects the value of all in an amount equal to the proportion of the amount of increase or decrease to the sum of the two.

2. To that part of criticism marked "B"—

Relative to the part of your criticism which I have marked "B," you state: "What is each note or bill of exchange but a promise to pay the legal money? *Its* value (the note or bill) is only so many of the standard units. If those units have been diminished in value, *all* your notes and bills have suffered the same deterioration."

I agree fully with you that the debasing of the

standard debases or reduces in value outstanding forms of credit circulation or substitutes for money. Where we differ is as to what must be considered in determining how much the "standard units" have been diminished in value. My contention is that where credit substitutes for the "standard units" do the work in redeeming these notes or bills of exchange, as well as "standard units" themselves would do it, under the law of supply and demand the amount of credit substitutes in existence affect the value of the standard units, and therefore the value of these units is determined by the relation of the factor of debasement to the sum of "standard units" and credit substitutes, not simply, as you suggest, upon the relation of the factor of debasement to the amount of "standard units" alone.

In order to make myself clear I accept your illustration as to wheat. You say: "As a question of exchange it seems to me to be essentially the same as if your notes and bills were payable in another commodity, say wheat. If wheat has gone down from $1 to 50 cents a bushel during the life of your paper, the makers will only pay half as much to take it up. It is this relation of all *promises to pay* to the *value* of the *unit* in which *payment is to be made*, which seems to me to make it fatally erroneous to treat these promises to pay as if 'real money.' In 'every corner' of a market the tremendous nature of the transaction and its consequences is determined by the prices of the *article*, though there may be very few bushels actually in the elevators, etc."

Now, in my following illustration, wheat may be regarded as gold, and corn as credit substitutes for gold. Suppose the community, as you suggest, made its notes and bills payable in the commodity wheat. Suppose, however, that through a new system of grinding in a mill called a bank, corn was made equally as desirable to the people as a medium of exchange as wheat — so desirable that, though not legally bound to do so, they would accept payments of their contracts in corn as readily as in wheat, knowing that, if they desired, they could turn corn into wheat "upon demand" at the same mill. It is evident, I think, under such relations, that the fluctuations in the supply of corn in existence would affect the price of wheat, and

6

that it would be the total amount of wheat and corn in existence which would be the chief factor in determining the ease or difficulty with which contracts payable " in wheat " were complied with. No business man, under such conditions, would attempt to forecast the price of wheat without fully considering the situation of the corn supply. While, as you suggest, the debasement in the price of wheat from $1 to 50 cents is quite a benefit to the makers of contracts payable in wheat, yet that debasement could occur either from a great increase in the supply of wheat, or a great increase in the supply of corn, or a great increase in the supply of both together.

In the light of this argument I will qualify the following proposition in your letter: " This seems to be the heart and soul of the whole theory of a standard. If, by any means, natural or artificial, the value of the standard is changed, you affect in the same ratio the value of all outstanding contracts to pay in that standard."

My qualification is that where, as under our present system, a promise to pay in the standard, called a check, is accepted as readily as the stand-

ard itself, in payment of another promise to pay
called a promissory note, or as purchase price of
a commodity, the amount of checks (or substitutes
for the standard) have more of an effect on the
value of the standard (since they are more
used than the standard) than the amount of
the standard has upon the value of the checks.
And as regards promissory notes, the ease or
difficulty with which they are paid, or rather the
amount of commodities at a given time necessary
to pay them, is determined by the sum of the
standard, and of circulating substitutes for the
standard, whether these substitutes be issued
by the Government or through the operations of
banks.